Student Edition

Eureka Math
Grade K
Modules 2 & 3

Special thanks go to the Gordon A. Cain Center and to the Department of Mathematics at Louisiana State University for their support in the development of *Eureka Math* .

For a free *Eureka Math* Teacher Resource Pack, Parent Tip Sheets, and more please visit www.Eureka.tools

Name _____ Leo _____   Date _____

Draw a line from the shape to its matching object.

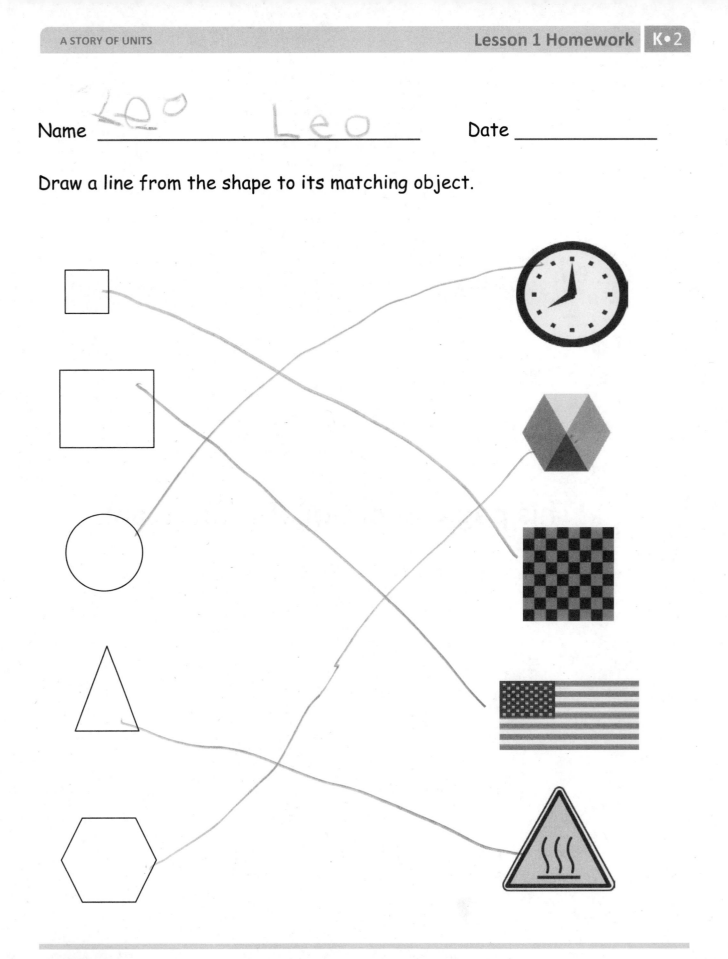

Lesson 1:   Find and describe flat triangles, squares, rectangles, hexagons, and circles using informal language without naming.

©2015 Great Minds. eureka-math.org
GK-M2-SE-B2-1.3.1-01.2016

5

This page intentionally left blank

Name __Leonel__   Date __Leonel__

Find the triangles, and color them blue. Put an X on shapes that are not triangles.

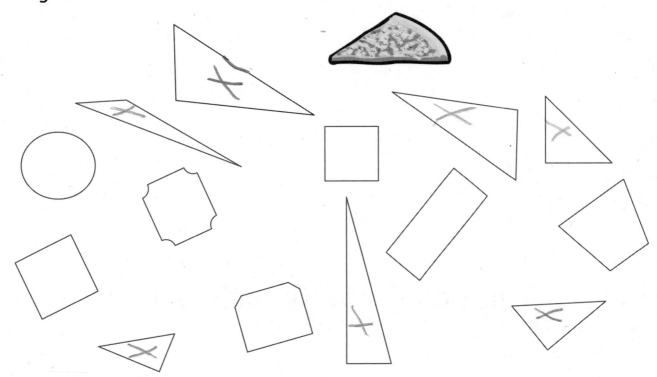

Draw some triangles.

EUREKA
MATH™

Lesson 2:   Explain decisions about classifications of triangles into categories using variants and non-examples. Identify shapes as triangles.

7

©2015 Great Minds. eureka-math.org
GK-M2-SE-B2-1.3.1-01.2016

This page intentionally left blank

Name _Leonel_____   Date _____

Color the triangles red and the other shapes blue.

Draw 2 different triangles of your own.

Lesson 2:   Explain decisions about classifications of triangles into categories using
variants and non-examples. Identify shapes as triangles.

9

This page intentionally left blank

Name _____   Date _____

Color all the rectangles red. Color all the triangles green.

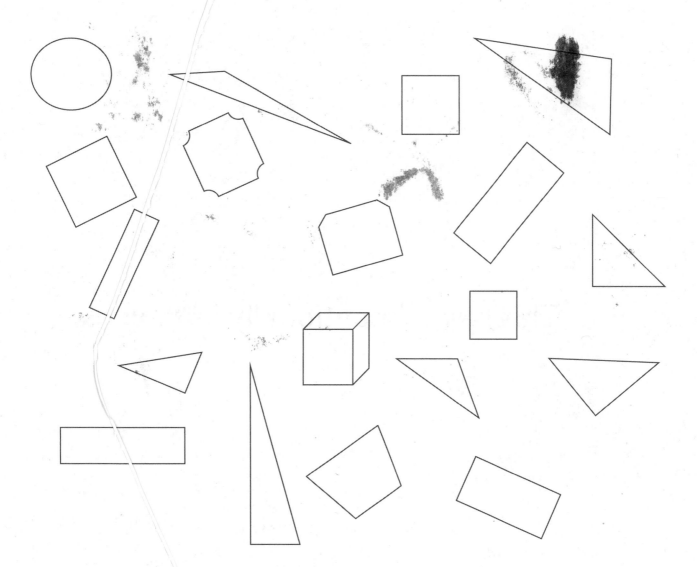

On the back of your paper, draw 2 rectangles and 3 triangles.

How many shapes did you draw? Put your answer in the circle.

EUREKA
MATH™

Lesson 3:   Explain decisions about classifications of rectangles into categories
using variants and non-examples. Identify shapes as rectangles.

13

©2015 Great Minds. eureka-math.org
GK-M2-SE-B2-1.3.1-01.2016

This page intentionally left blank

Name _Leonel_____   Date _____

EUREKA MATH   **Lesson 5:**   Describe and communicate positions of all flat shapes using the words   **23**
                                    above, below, beside, in front of, next to, and *behind.*

©2015 Great Minds. eureka-math.org
GK-M2-SE-B2-1.3.1-01.2016

This page intentionally left blank

Name Leonel _____   Date _____

- **Behind** the elephant, draw a shape with 4 straight sides that are exactly the same length. Color it blue.

- **Above** the elephant, draw a shape with no corners. Color it yellow.

- **In front of** the elephant, draw a shape with 3 straight sides. Color it green.

- **Below** the elephant, draw a shape with 4 sides, 2 long and 2 short. Color it red.

- **Below** the elephant, draw a shape with 6 corners. Color it orange.

On the back of your paper, draw 1 hexagon and 4 triangles.
How many shapes did you draw? Put your answer in the circle.

      Lesson 5:   Describe and communicate positions of all flat shapes using the words
                          above, below, beside, in front of, next to, and *behind.*   25

©2015 Great Minds. eureka-math.org
GK-M2-SE-B2-1.3.1-01.2016

This page intentionally left blank

Name __Leonel__   Date _____

Find things in your house or in a magazine that look like these solids. Draw the solids or cut out and paste pictures from a magazine.

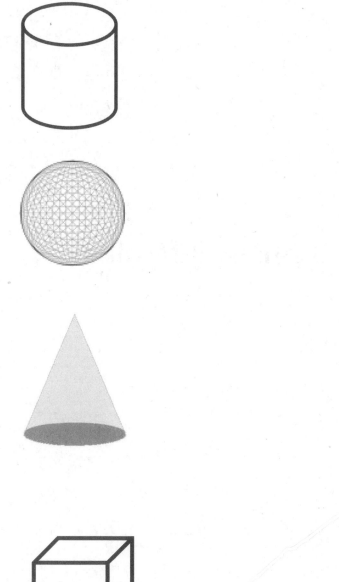

Lesson 6:   Find and describe solid shapes using informal language without naming.

29

©2015 Great Minds. eureka-math.org
GK-M2-SE-B2-1.3.1-01.2016

This page intentionally left blank

Name _____   Date _____

Cut one set of solid shapes. Sort the 4 solid shapes. Paste onto the chart.

These have corners.   These do not have corners.

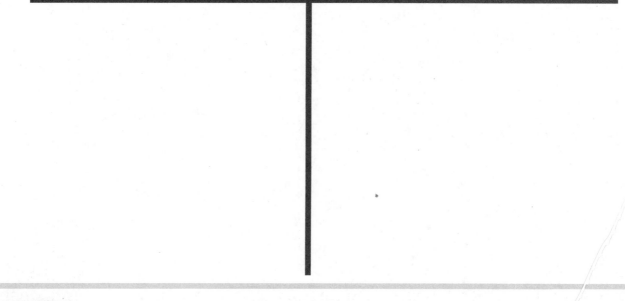

Cut the other set of solid shapes, and make a rule for how you sorted them. Paste onto the chart.

EUREKA MATH

Lesson 7:   Explain decisions about classification of solid shapes into categories.
Name the solid shapes.

©2015 Great Minds. eureka-math.org
GK-M2-SE-B2-1.3.1-01.2016

33

This page intentionally left blank

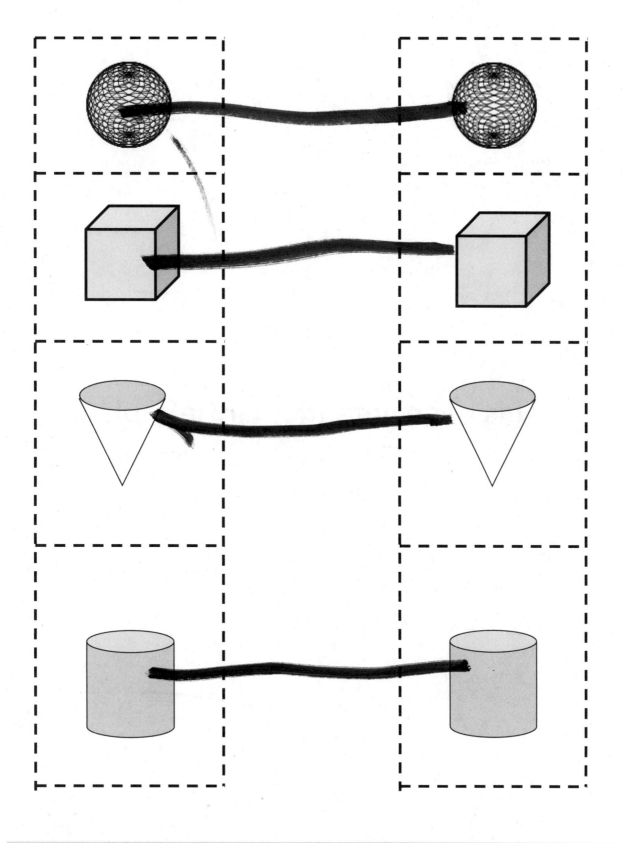

Lesson 7:   Explain decisions about classification of solid shapes into categories.
Name the solid shapes.

35

©2015 Great Minds. eureka-math.org
GK-M2-SE-B2-1.3.1-01.2016

This page intentionally left blank

Name _____ Date _____

Tell someone at home the names of each solid shape.

   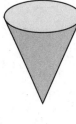 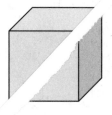

Sphere   Cylinder   Cone   Cube

Color the car **beside** the stop sign green.

Circle the **next** car with blue.

Color the car **behind** the circled car red.

Draw a road **below** the cars.

Draw a policeman **in front of** the cars.

Draw a sun **above** the cars.

Lesson 8: Describe and communicate positions of all solid shapes using the words *above, below, beside, in front of, next to,* and *behind.*

41

©2015 Great Minds. eureka-math.org
GK-M2-SE-B2-1.3.1-01.2016

This page intentionally left blank

Name _____   Date _____

Circle the pictures of the flat shapes with red. Circle the pictures of the solid shapes with green.

Lesson 9:   Identify and sort shapes as two-dimensional or three-dimensional, and recognize two-dimensional and three-dimensional shapes in different orientations and sizes.

43

©2015 Great Minds. eureka-math.org
GK-M2-SE-B2-1.3.1-01.2016

This page intentionally left blank

Name _____   Date _____

In each row, circle the one that doesn't belong. Explain your choice to a grown-up.

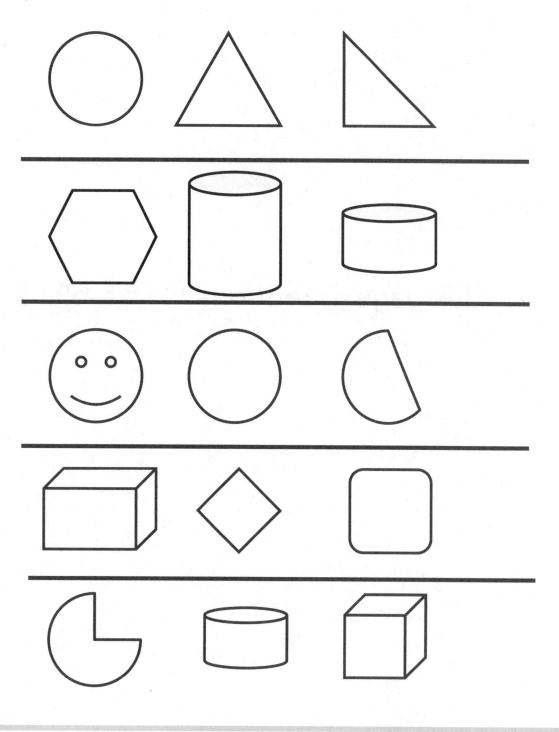

Lesson 9:   Identify and sort shapes as two-dimensional or three-dimensional, and recognize two-dimensional and three-dimensional shapes in different orientations and sizes.

45

©2015 Great Minds. eureka-math.org
GK-M2-SE-B2-1.3.1-01.2016

This page intentionally left blank

Name _____   Date _____

Shape Up Your Kitchen!

Search your kitchen to see what shapes and solids you can find. Make a kitchen-shaped collage by drawing the shapes that you see and by tracing the faces of the solids that you find. Color your collage.

EUREKA MATH

Lesson 10:   Culminating task—collaborative groups create displays of different flat shapes with examples, non-examples, and a corresponding solid shape.

©2015 Great Minds. eureka-math.org
GK-M2-SE-B2-1.3.1-01.2016

47

This page intentionally left blank

Name _____   Date _____

These are (____).   These are not (____).

_____
work mat

Lesson 10:   Culminating task—collaborative groups create displays of different flat shapes with examples, non-examples, and a corresponding solid shape.

49

©2015 Great Minds. eureka-math.org
GK-M2-SE-B2-1.3.1-01.2016

This page intentionally left blank

Eureka Math
Grade K
Module 3

Special thanks go to the Gordon A. Cain Center and to the Department of Mathematics at Louisiana State University for their support in the development of *Eureka Math* .

Name _____   Date _____

For each pair, circle the longer one. Imagine the paper strips are lying flat on a table.

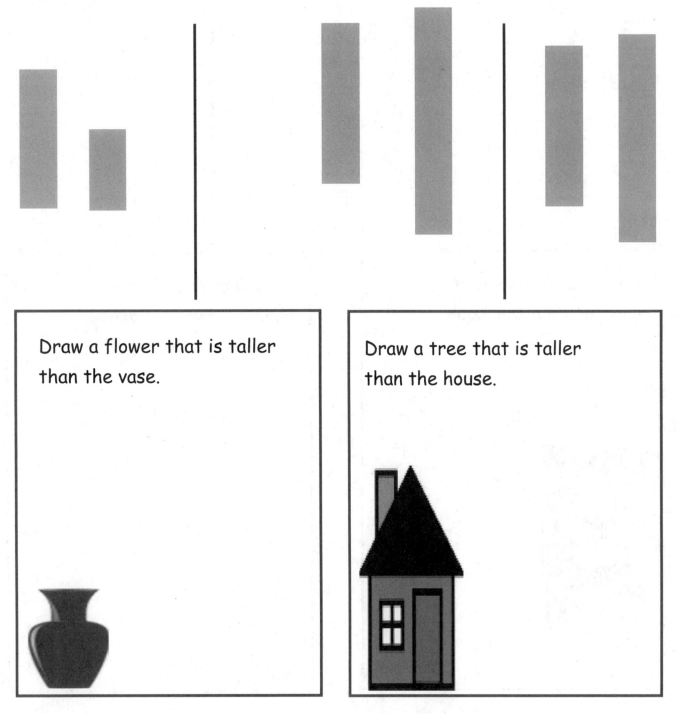

Draw a flower that is taller than the vase.

Draw a tree that is taller than the house.

Lesson 1:   Compare lengths using *taller than* and *shorter than* with aligned and non-aligned endpoints.

©2015 Great Minds. eureka-math.org
GK-M3-SE-B2-1.3.1-01.2016

1

For each pair, circle the shorter one.

Draw a bookmark that is shorter than this book.

Draw a crayon that is shorter than this pencil.

Lesson 1:   Compare lengths using *taller than* and *shorter than* with aligned and non-aligned endpoints.

Name _____   Date _____

Draw 3 more flowers that are shorter than these flowers.
Count how many flowers you have now. Write the number in
the box.

Draw 2 more ladybugs that are taller than these ladybugs.
Count how many ladybugs you have now. Write the number in
the box.

On the back of your paper, draw something that is taller than you.
Draw something that is shorter than a flagpole.

Lesson 1:   Compare lengths using *taller than* and *shorter than* with aligned and
                   non-aligned endpoints.

3

©2015 Great Minds. eureka-math.org
GK-M3-SE-B2-1.3.1-01.2016

This page intentionally left blank

Name _____   Date _____

Cut out the picture of the string at the bottom of the page. Compare the string with each object to see which is longer. Use the line next to each object to help you compare. Color objects shorter than the string green. Color objects longer than the string orange.

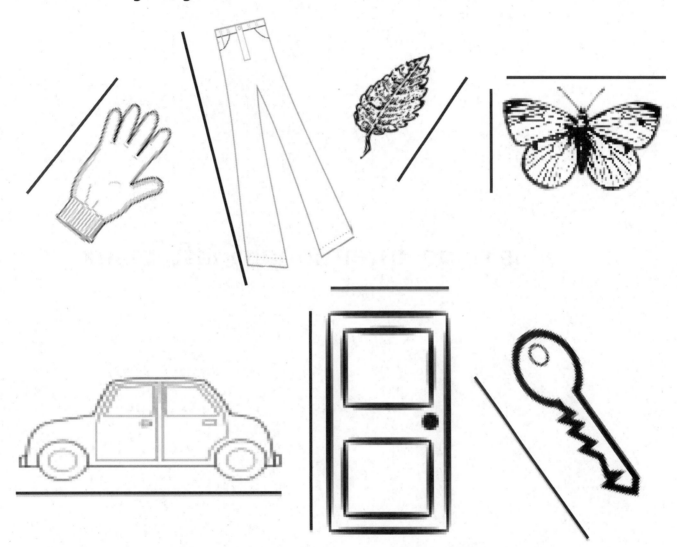

On the back of your paper, draw something longer than, shorter than, and the same length as the picture of the string. Color objects shorter than the string green. Color objects longer than the string orange.

©2015 Great Minds. eureka-math.org
GK-M3-SE-B2-1.3.1-01.2016

This page intentionally left blank

Name _____   Date _____

Using the piece of string from class, find three items at home that are shorter than your piece of string and three items that are longer than your piece of string. Draw a picture of those objects on the chart. Try to find at least one thing that is about the same length as your string, and draw a picture of it on the back.

Shorter than the string	Longer than the string

This page intentionally left blank

Longer or Shorter Recording Sheet

These objects are **longer than** my string.	These objects are **shorter than** my string.

_____

longer or shorter

This page intentionally left blank

Directions: Pretend that I am a pirate who has traveled far away from home. I miss my house and family. Will you draw a picture as I describe my home? Listen carefully, and draw what you hear.

- Draw a house in the middle of the paper as tall as your pointer finger.

- Now, draw my daughter. She is shorter than the house.

- There's a great tree in my yard. My daughter and I love to climb the tree. The tree is taller than my house.

- My daughter planted a beautiful daisy in the yard. Draw a daisy that is shorter than my daughter.

- Draw a branch lying on the ground in front of the house. Make it the same length as the house.

- Draw a caterpillar next to the branch. My parrot loves to eat caterpillars. Of course, the length of the caterpillar is shorter than the length of the branch.

- My parrot is always hungry, and there are plenty of bugs for him to eat at home. Draw a ladybug above the caterpillar. Should the ladybug be shorter or longer than the branch?

- Now, draw some more things you think my family would enjoy.

Show your picture to your partner, and talk about the extra things that you drew. Use *longer than* and *shorter than* when you are describing them.

This page intentionally left blank

Name _____   Date _____

Home is where the heARRt is, matey.

EUREKA
MATH™

Lesson 3:   Make a series of *longer than* and *shorter than* comparisons.

13

This page intentionally left blank

Name _____ Date _____

Take out a new crayon. Circle objects with lengths shorter than the crayon blue. Circle objects with lengths longer than the crayon red.

On the back of your paper, draw some things shorter than the crayon and longer than the crayon. Draw something that is the same length as the crayon.

This page intentionally left blank

Longer than...

Shorter than...

_____

longer than and shorter than work mat

©2015 Great Minds. eureka-math.org
GK-M3-SE-B2-1.3.1-01.2016

This page intentionally left blank

Name _____ Date _____

Circle the shorter stick. The squares represent linking cubes.

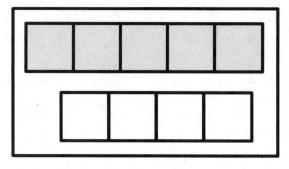

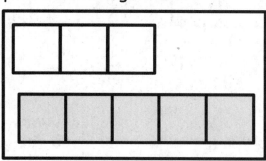

How many linking cubes are in the shorter stick?

How many linking cubes are in the shorter stick?

Circle the longer stick.

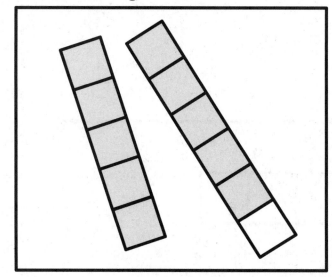

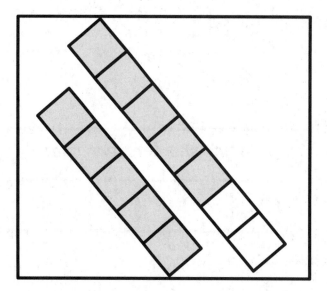

How many linking cubes are in the longer stick?

How many linking cubes are in the longer stick?

Lesson 4:   Compare the length of linking cube sticks to a 5-stick.

19

©2015 Great Minds. eureka-math.org
GK-M3-SE-B2-1.3.1-01.2016

Draw a stick **shorter than** my 5-stick.

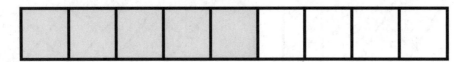

Draw a stick **longer than** mine.

Draw a stick **shorter than** mine.

Lesson 4:   Compare the length of linking cube sticks to a 5-stick.

Name _____   Date _____

Use a red crayon to circle the sticks that are
shorter than the 5-stick.

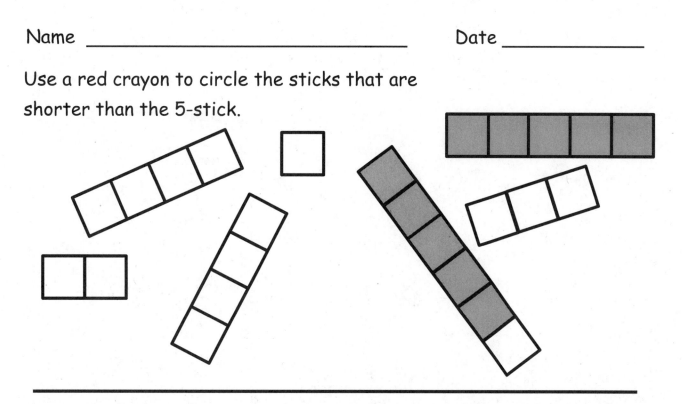

_____

Use a blue crayon to circle the sticks that are longer than the 5-stick.

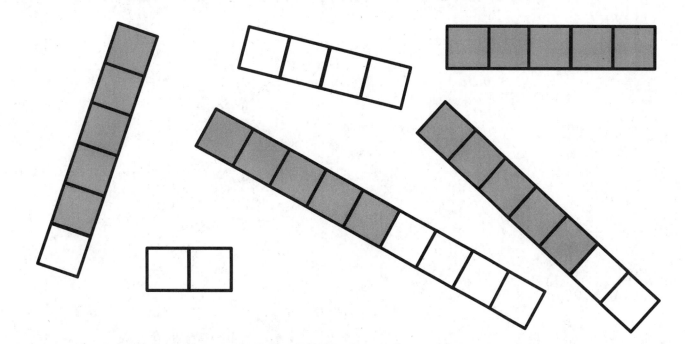

On the back, draw a 7-stick. Draw a stick longer than it. Draw a stick
shorter than it.

EUREKA
MATH™

Lesson 4:   Compare the length of linking cube sticks to a 5-stick.

21

©2015 Great Minds. eureka-math.org
GK-M3-SE-B2-1.3.1-01.2016

This page intentionally left blank

Longer than my 5-stick:

Shorter than my 5-stick:

longer or shorter mat

Lesson 4:   Compare the length of linking cube sticks to a 5-stick.

23

©2015 Great Minds. eureka-math.org
GK-M3-SE-B2-1.3.1-01.2016

This page intentionally left blank

Name _____   Date _____

Circle the stick that is shorter than the other.

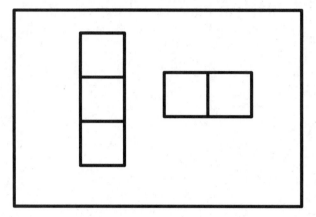   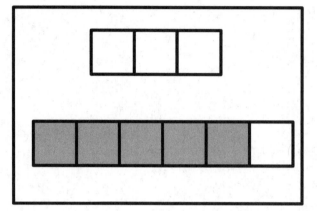

Circle the stick that is longer than the other.

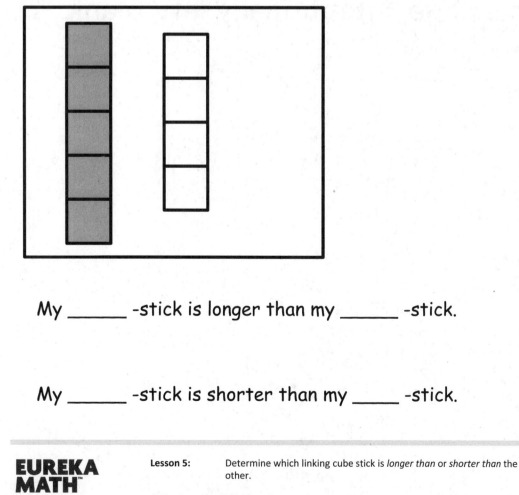

My _____ -stick is longer than my _____ -stick.

My _____ -stick is shorter than my _____ -stick.

This page intentionally left blank

Circle the stick that is shorter than the other stick.

My _____ -stick is longer than my _____ -stick.

My _____ -stick is shorter than my _____ -stick.

On the back of your paper, draw a 6-stick.

Draw a stick longer than your 6-stick.

Draw a stick shorter than your 6-stick.

OR

On the back of your paper, draw a 9-stick.

Draw a stick longer than your 9-stick.

Draw a stick shorter than your 9-stick.

This page intentionally left blank

Name _____   Date _____

Circle the stick that is shorter than the other.

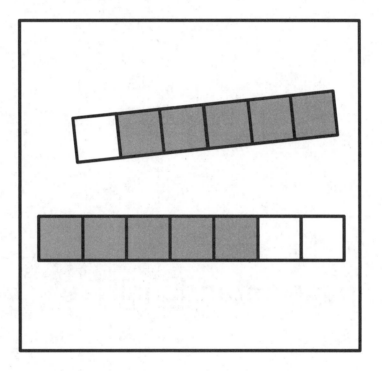

My _____ -stick is shorter than my _____ -stick.

My _____ -stick is longer than my _____ -stick.

On the back of your paper, draw a 7-stick.

Draw a stick that is longer than the 7-stick.

Draw a stick that is shorter than the 7-stick.

Lesson 5:   Determine which linking cube stick is *longer than* or *shorter than* the other.

29

This page intentionally left blank

Circle the stick that is longer than the other.

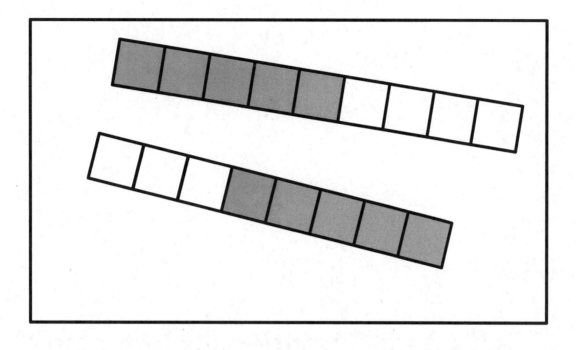

My _____ -stick is shorter than my _____ -stick.

My _____ -stick is longer than my _____ -stick.

On the back of your paper, draw a stick that is between a 4- and a 6-stick.

Draw a stick that is longer than your new stick.

Draw a stick this is shorter than your new stick.

Lesson 5:   Determine which linking cube stick is *longer than* or *shorter than* the other.

©2015 Great Minds. eureka-math.org
GK-M3-SE-B2-1.3.1-01.2016

31

This page intentionally left blank

Name _____   Date _____

In the box, write the number of linking cubes there are in the stick.
Draw a green circle around the stick if it is longer than the object.
Draw a blue circle around the stick if it is shorter than the object.

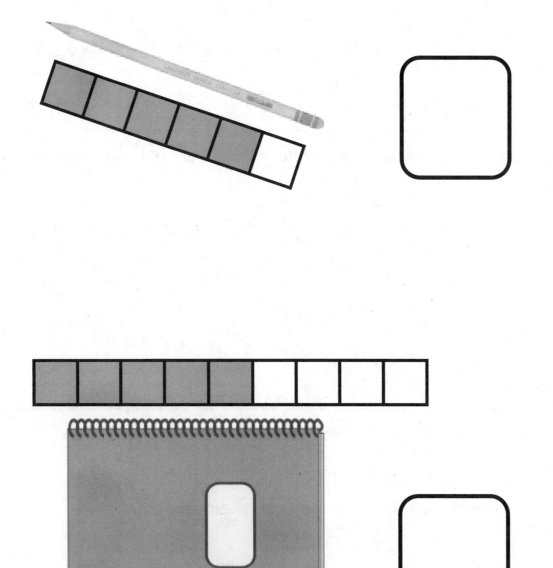

Make a 3-stick. In your classroom, select a crayon, and see if your crayon is longer than or shorter than your stick.

Trace your 3-stick and your crayon to compare their lengths.

In your classroom, find a marker, and make a stick that is longer than your marker.

Trace your stick and your marker to compare their lengths.

Make a 5-stick. Find something in the classroom that is longer than your 5-stick.

Trace your 5-stick and the object to compare their lengths.

Lesson 6:   Compare the length of linking cube sticks to various objects.

Name _____   Date _____

Color the cubes to show the length of the object.

Lesson 6:   Compare the length of linking cube sticks to various objects.

35

This page intentionally left blank

Name _____   Date _____

These boxes represent cubes.

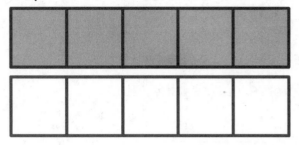

Color 2 cubes green. Color 3 cubes blue.

Together, my green 2-stick and blue 3-stick are the same length as 5 cubes.

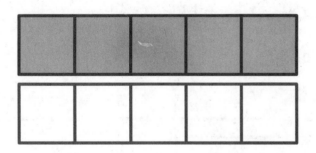

Color 3 cubes blue. Color 2 cubes green.

Together, my blue 3-stick and green 2-stick are the same length as

___ cubes.

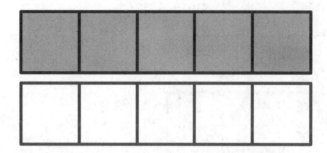

Color 1 cube green. Color 4 cubes blue.

How many did you color? _____

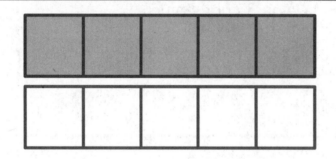

Color 4 cubes green. Color 1 cube blue.

How many did you color? _____

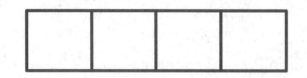

Color 2 cubes yellow. Color 2 cubes blue.

Together, my 2 yellow and 2 blue are the same as _____.

©2015 Great Minds. eureka-math.org
GK-M3-SE-B2-1.3.1-01.2016

My 5:

My _____:

My _____:

_____

riddle work mat

My
5:

My
_____:

My
_____:

riddle work mat

Lesson 7:   Compare objects using *the same as*.

©2015 Great Minds. eureka-math.org
GK-M3-SE-B2-1.3.1-01.2016

Name _____   Date _____

Draw an object that would be lighter than the one in the picture.

This page intentionally left blank

Name _____ Date _____

Draw something inside the box that is heavier than the object on the balance.

Lesson 9:   Compare objects using *heavier than, lighter than,* and *the same as* with balance scales.

©2015 Great Minds. eureka-math.org
GK-M3-SE-B2-1.3.1-01.2016

47

Draw something lighter than the object on the balance.

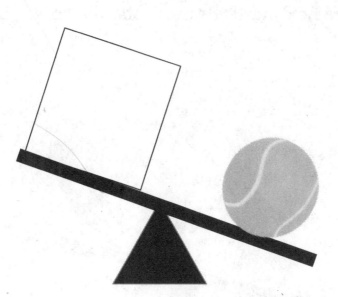

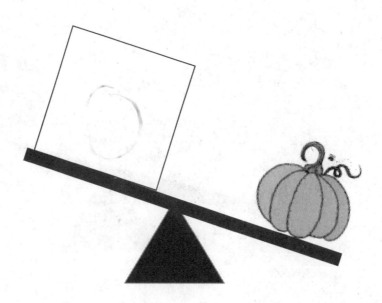

Lesson 9:   Compare objects using *heavier than, lighter than*, and *the same as* with balance scales.

Name _____ Date _____

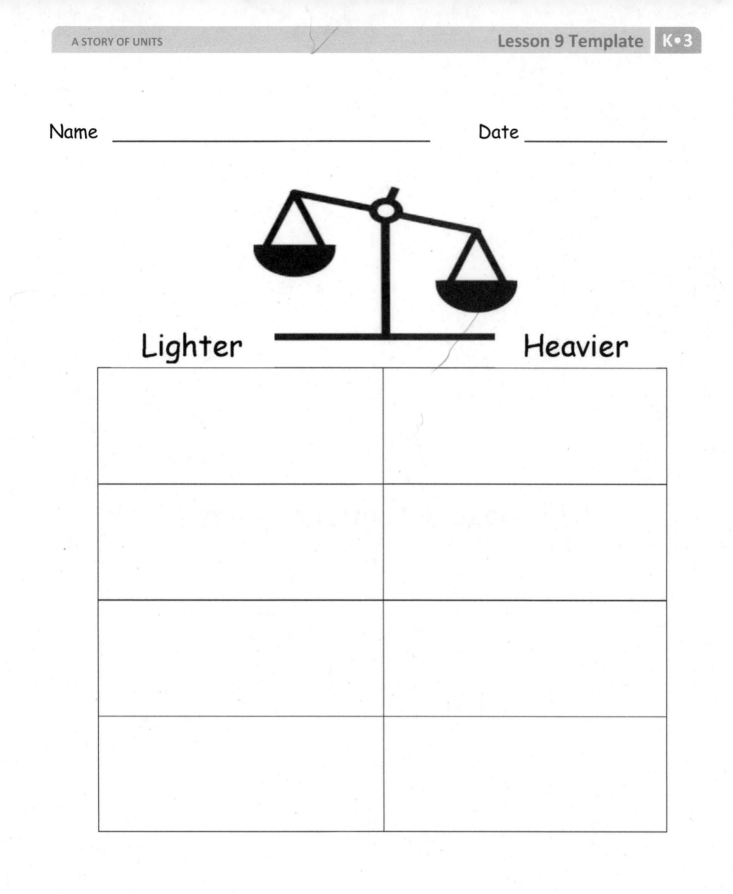

Lighter   Heavier

_____

lighter or heavier recording sheet

Lesson 9: Compare objects using *heavier than, lighter than,* and *the same as* with balance scales.

49

This page intentionally left blank

Name _____   Date _____

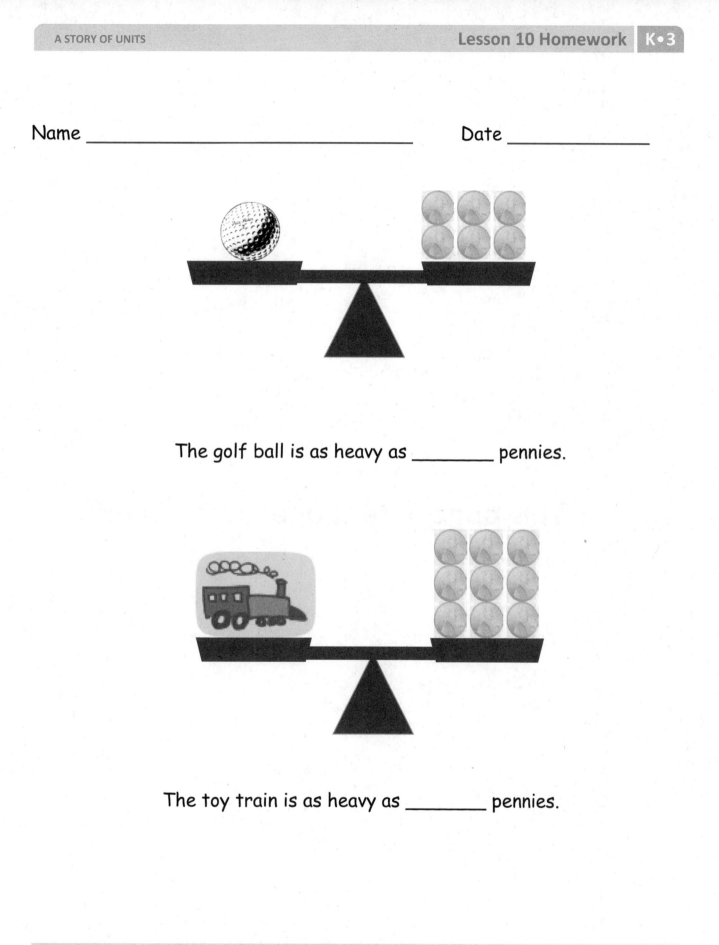

The golf ball is as heavy as _____ pennies.

The toy train is as heavy as _____ pennies.

Lesson 10:   Compare the weight of an object to a set of unit weights on a balance
scale.

51

This page intentionally left blank

Draw in the pennies so the carrot is as heavy as 5 pennies.

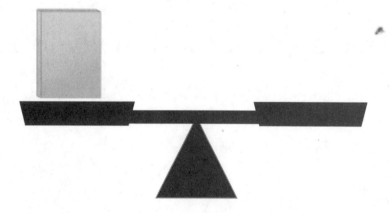

Draw in the pennies so the book is as heavy as 10 pennies.

On the back of your paper, draw a balance scale with an object. Write how many pennies you think the object would weigh. If you can, bring in the object tomorrow. We will weigh it to see if it weighs as many pennies as you thought.

Lesson 10:   Compare the weight of an object to a set of unit weights on a balance scale.

53

©2015 Great Minds. eureka-math.org
GK-M3-SE-B2-1.3.1-01.2016

This page intentionally left blank

Name _____   Date _____

is as heavy as _____ pennies.

is as heavy as _____ pennies.

is as heavy as _____ pennies.

is as heavy as _____ pennies.

_____

as heavy as recording sheet

This page intentionally left blank

Name L Leonel _____   Date _____

Draw a line from the balance to the linking cubes that weigh the same.

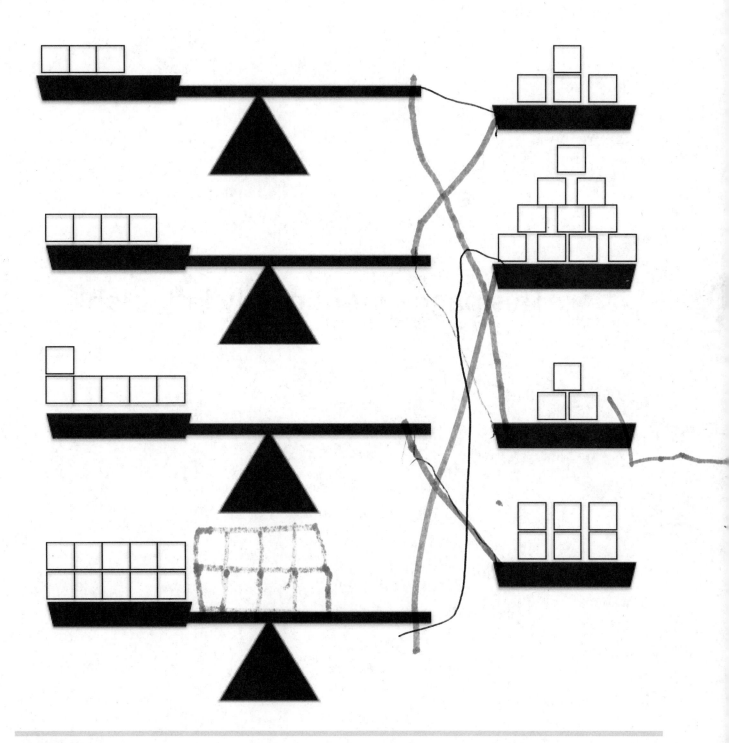

Lesson 11:   Observe conservation of weight on the balance scale.

57

This page intentionally left blank

Name __Leonel_____   Date _____

Draw linking cubes so each side weighs the same.

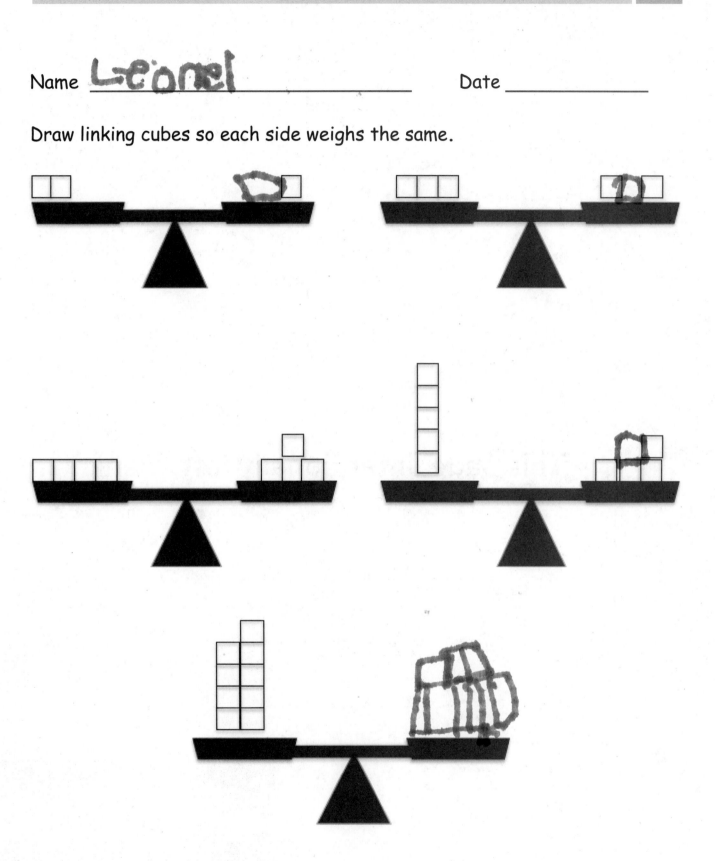

EUREKA
MATH™

Lesson 11:   Observe conservation of weight on the balance scale.

59

©2015 Great Minds. eureka-math.org
GK-M3-SE-B2-1.3.1-01.2016

This page intentionally left blank

Name _____   Date _____

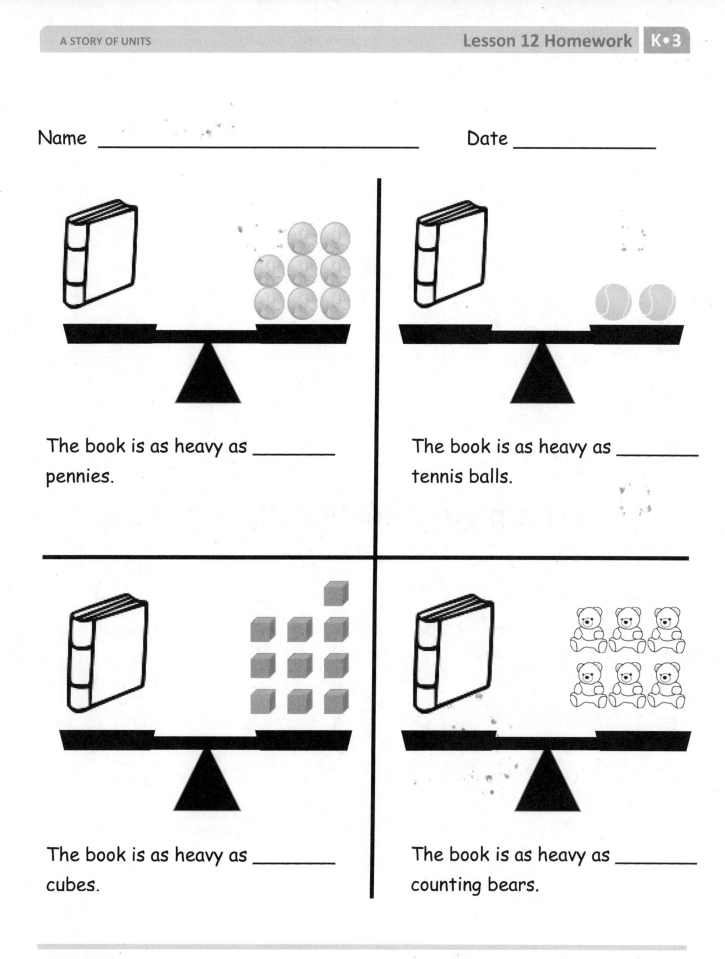

The book is as heavy as _____ pennies.

The book is as heavy as _____ tennis balls.

The book is as heavy as _____ cubes.

The book is as heavy as _____ counting bears.

Lesson 12:   Compare the weight of an object with sets of different objects on a balance scale.

61

©2015 Great Minds. eureka-math.org
GK-M3-SE-B2-1.3.1-01.2016

This page intentionally left blank

Name _____   Date _____

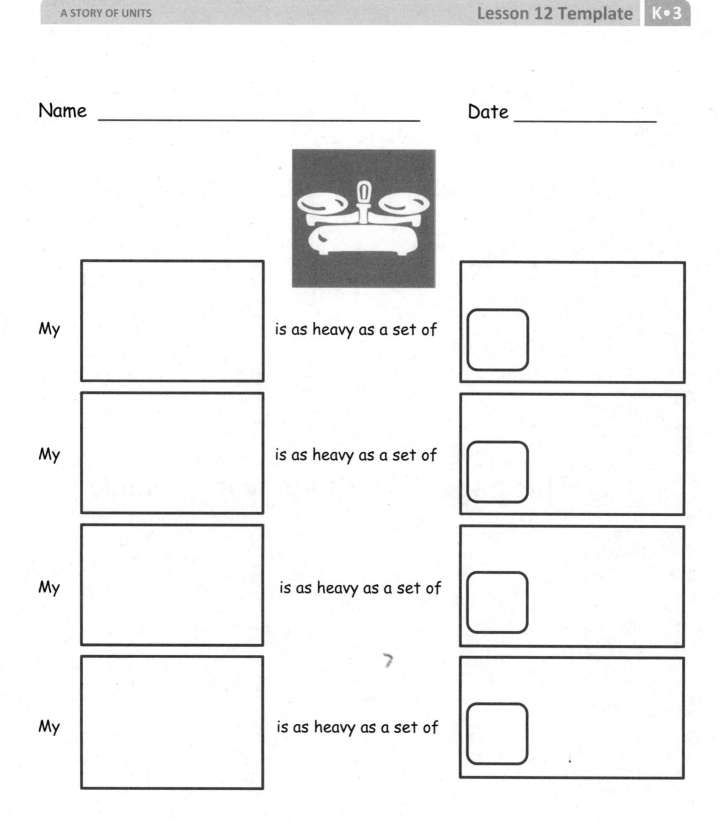

My [ ]   is as heavy as a set of   [ ]

My [ ]   is as heavy as a set of   [ ]

My [ ]   is as heavy as a set of   [ ]

My [ ]   is as heavy as a set of   [ ]

_____

as heavy as a set recording sheet

Lesson 12:   Compare the weight of an object with sets of different objects on a balance scale.

63

©2015 Great Minds. eureka-math.org
GK-M3-SE-B2-1.3.1-01.2016

This page intentionally left blank

Name _____   Date _____

Talk to your partner about which container might have more or less capacity. Which might have about the same capacity? What happens if the containers are not filled up to the top? Can we tell that they are filled completely from looking at the pictures?

EUREKA
MATH™

Lesson 13:   Compare volume using *more than, less than,* and *the same as* by
pouring.

©2015 Great Minds. eureka-math.org
GK-M3-SE-B2-1.3.1-01.2016

65

This page intentionally left blank

Name _____   Date _____

In class, we have been working on capacity. Encourage your child to explore with different-sized containers to see which ones have the biggest and smallest capacity. Children can experiment by pouring liquid from one container to another.

All the homework you will see for the next few days will be a review of skills taught from Module 1.

Each rectangle shows 6 objects. Circle 2 different sets within each. The two sets represent the two parts that make up the 6 objects. The first one has been done for you.

Lesson 13:   Compare volume using *more than, less than,* and *the same as* by pouring.

67

©2015 Great Minds. eureka-math.org
GK-M3-SE-B2-1.3.1-01.2016

This page intentionally left blank

Name _____   Date _____

I found out that this container held the most rice.

It had the biggest capacity.

I found out that this container held the least rice.

It had the smallest capacity.

capacity recording sheet

EUREKA
MATH™

Lesson 13: Compare volume using *more than, less than,* and *the same as* by pouring.

69

©2015 Great Minds. eureka-math.org
GK-M3-SE-B2-1.3.1-01.2016

This page intentionally left blank

Name _Leonel_____   Date _____

Within each rectangle, make one set of 6 objects. The first one has been done for you.

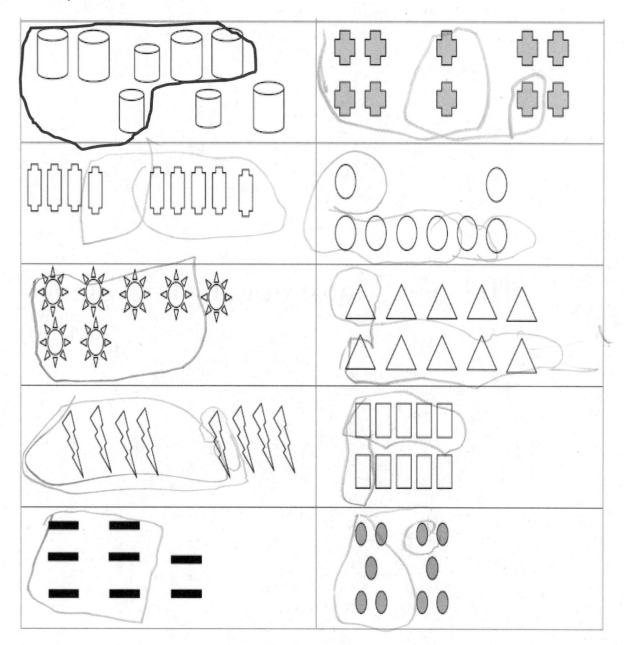

This page intentionally left blank

Name _____   Date _____

My cup of rice looks like:

Now it looks like:

Now it looks like:

Now it looks like:

volume recording sheet

This page intentionally left blank

Name _____   Date _____

Circle 2 sets within each set of 7. The first one has been done for you.

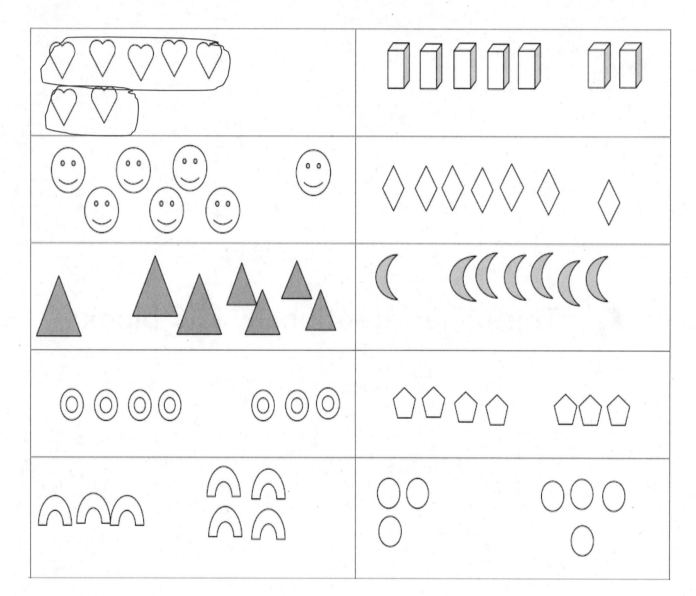

This page intentionally left blank

Name _____   Date _____

We've Got the Scoop!

is the same as _____ scoops.

is the same as _____ scoops.

is the same as _____ scoops.

_____ scoops is the same as

_____ scoops is the same as

we've got the scoop recording sheet

This page intentionally left blank

Name _____   Date _____

Cover the shape with squares. Count how many, and write the number in the box.

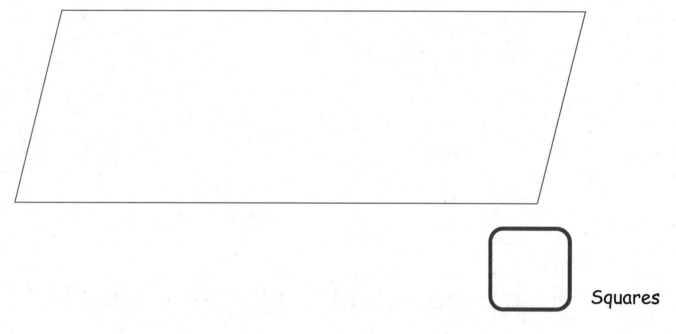

Squares

Cover the shape with beans. Count how many, and write the number in the box.

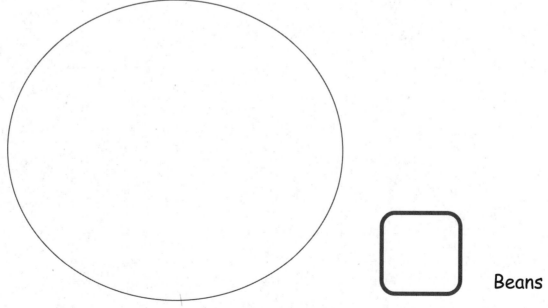

Beans

This page intentionally left blank

Name _____   Date _____

Trace your hand. Cover the tracing with pennies. Have an adult trace his
or her hand. Cover the tracing with pennies.* Whose hand is bigger?
How do you know that?

*Note: Instead of pennies, you can use pasta, beans, buttons, or
another coin. You may want to do this activity twice using different
materials to cover the hands. Talk about which materials took more or
less to cover and why.

This page intentionally left blank

Name _____   Date _____

My square.

My square covered
with a circle.

My square covered
with little squares.

My square covered
with beans.

_____
my square recording sheet

This page intentionally left blank

Name _____   Date _____

Draw straight lines with your ruler to see if there are enough flowers for the butterflies.

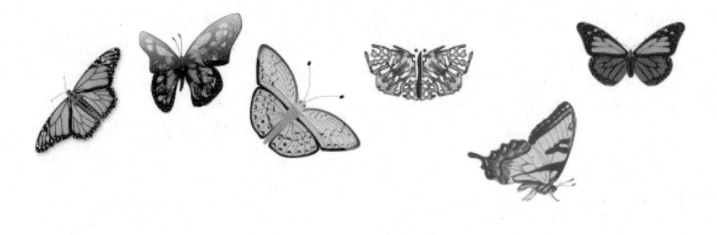

On the back, draw some plates. Draw enough apples so each plate has one.

EUREKA
MATH™

This page intentionally left blank

Name _____   Date _____

Draw straight lines with your ruler to see if there are enough shovels for the pails.

Make sure there is a fork for every plate. Draw straight lines with a ruler from each plate to a fork. If there are not enough forks, draw one.

You have 4 fishes. Draw enough fish bowls so you can put 1 fish in each fish bowl.

©2015 Great Minds. eureka-math.org
GK-M3-SE-B2-1.3.1-01.2016

Name _____   Date _____

Draw straight lines with your ruler to see if there are enough hats for the scarves.

Are there more [hat] or [scarf] ?

Cross off by putting an X on 2 [scarf] . Talk to your partner about what you notice now.

Draw more leaves than ants.

This page intentionally left blank

Name _____   Date _____

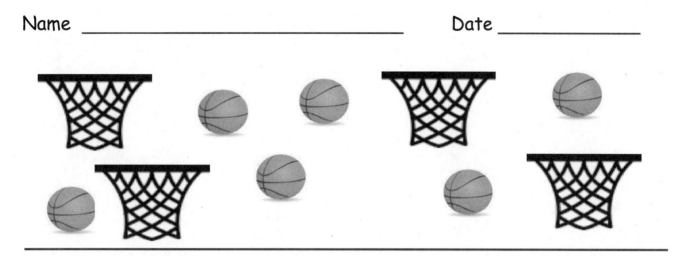

Draw straight lines with your ruler to see if there is one hoop for each ball.

Are there *more* 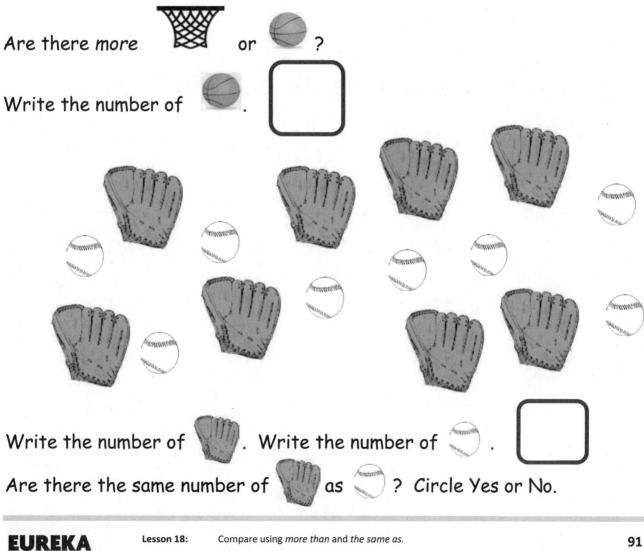 or ⚫ ?

Write the number of 🏀 . ⬜

Write the number of 🧤 . Write the number of ⚾ . ⬜

Are there the same number of 🧤 as ⚾ ? Circle Yes or No.

EUREKA
MATH™

This page intentionally left blank

Name _____   Date _____

Count the objects. Circle the set that has fewer.

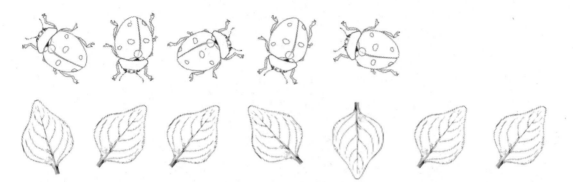

Draw more ladybugs so there are the same number of ladybugs as leaves.

Count the objects. Circle the set that has fewer.

Draw more watermelon slices so there are the same number of watermelon slices as peaches.

On the back, draw suns and stars. Draw fewer suns than stars.

Lesson 19:   Compare using *fewer than* and *the same as*.   93

©2015 Great Minds. eureka-math.org
GK-M3-SE-B2-1.3.1-01.2016

This page intentionally left blank

Name _____   Date _____

Draw another bird so there are the same number of birds as bird cages.

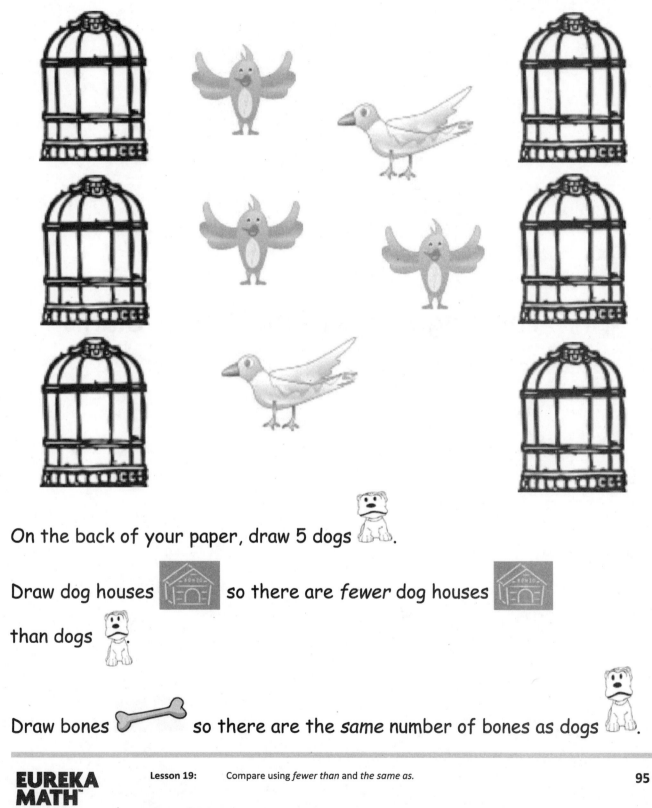

On the back of your paper, draw 5 dogs.

Draw dog houses so there are *fewer* dog houses than dogs.

Draw bones so there are the *same* number of bones as dogs.

This page intentionally left blank

Name _____   Date _____

Count the dots on the die. Color as many beads as the dots on the die.
Circle the longer chain in each pair.

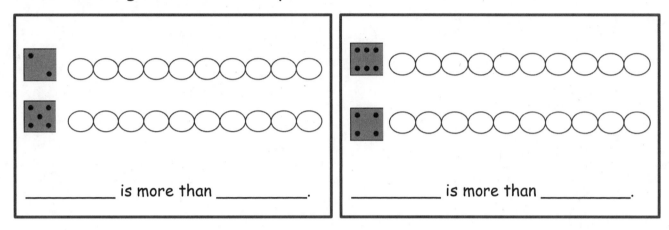

_____ is more than _____.   _____ is more than _____.

Roll the die. Write the number you roll in the box, and color that many
beads. Roll the die again, and do the same on the next set of beads. Circle
the chain with fewer beads.

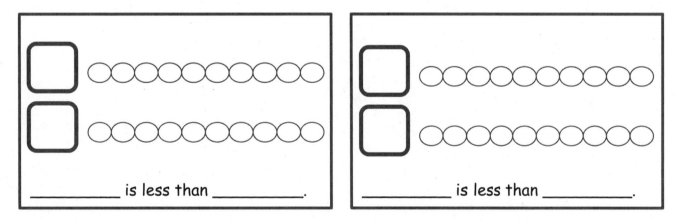

_____ is less than _____.   _____ is less than _____.

On the back, make more chains by rolling the die. Write the number you
rolled, and then make a chain with the same number you rolled.

This page intentionally left blank

Name _____   Date _____

On the first chain, color the first 3 beads blue.
On the next chain, color more than 3 beads red.
How many beads did you color red? Write the number in the box.

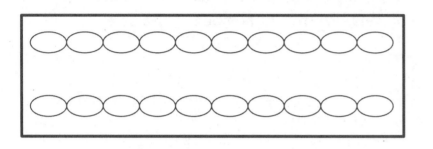

_____ red beads is more than 3.

On the first chain, color the first 5 beads green.
On the next chain, color fewer than 5 beads yellow.
How many beads did you color yellow? Write the number in the box.

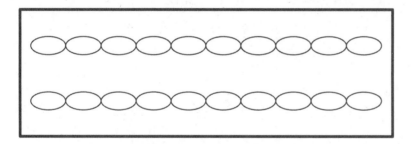

_____ yellow beads is fewer than 5.

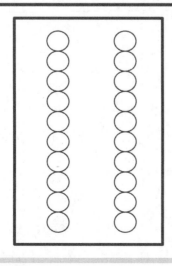

Color 2 beads brown in the first column.

Color more than 2 beads blue in the second column.

How many beads did you color in the second column?
Write the number in the box.

_____ blue beads is more than 2.

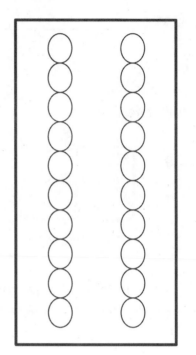

Color 9 beads red in the first column.

Color fewer than 9 beads green in the second column.

How many beads did you color in the second column? Write the number in the box.

_____ green beads is fewer than 9.

Draw a chain with more than 3 beads but fewer than 10 beads.

Draw a chain that has fewer than 10 beads but more than 4 beads.

Lesson 20:   Relate *more* and *less* to length.

©2015 Great Minds. eureka-math.org
GK-M3-SE-B2-1.3.1-01.2016

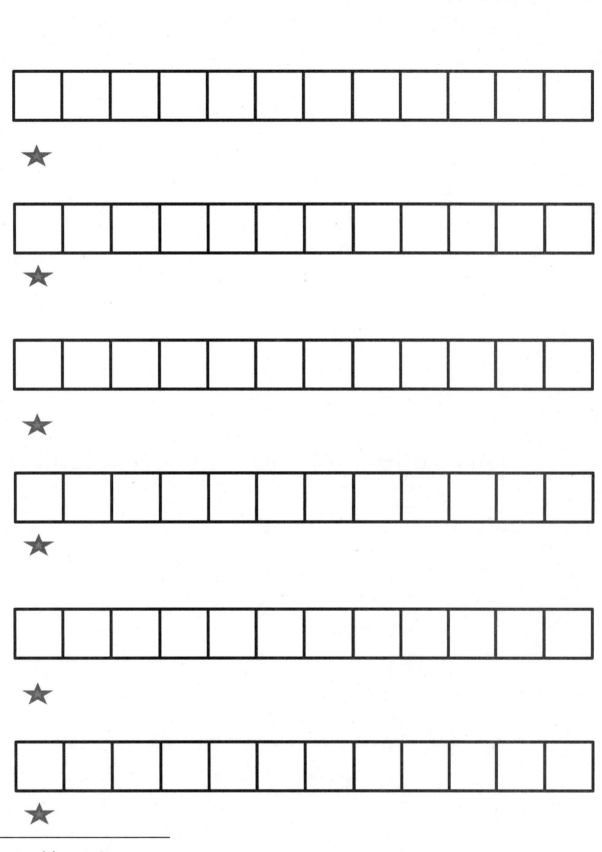

square path letter trains

©2015 Great Minds. eureka-math.org
GK-M3-SE-B2-1.3.1-01.2016

This page intentionally left blank

Name _____   Date _____

Which has more? The 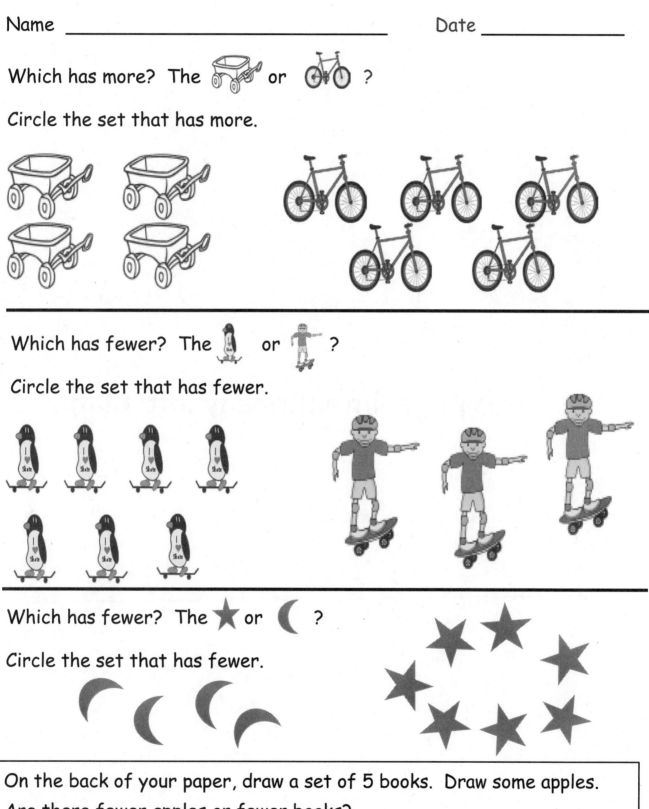 or ?

Circle the set that has more.

Which has fewer? The or ?

Circle the set that has fewer.

Which has fewer? The ★ or ☾ ?

Circle the set that has fewer.

On the back of your paper, draw a set of 5 books. Draw some apples. Are there fewer apples or fewer books?

This page intentionally left blank

Name _____   Date _____

Draw a shape to make the sentence true.

There are more _____than ⬡ .

There are fewer ▲ than _____.

There are fewer _____than ▬ .

_____

more than, fewer than recording sheet

EUREKA MATH™

Lesson 21:   Compare sets informally using *more*, *less*, and *fewer*.

107

This page intentionally left blank

Name _____   Date _____

Count the objects in the box. Then, draw the same number of circles in the empty box.

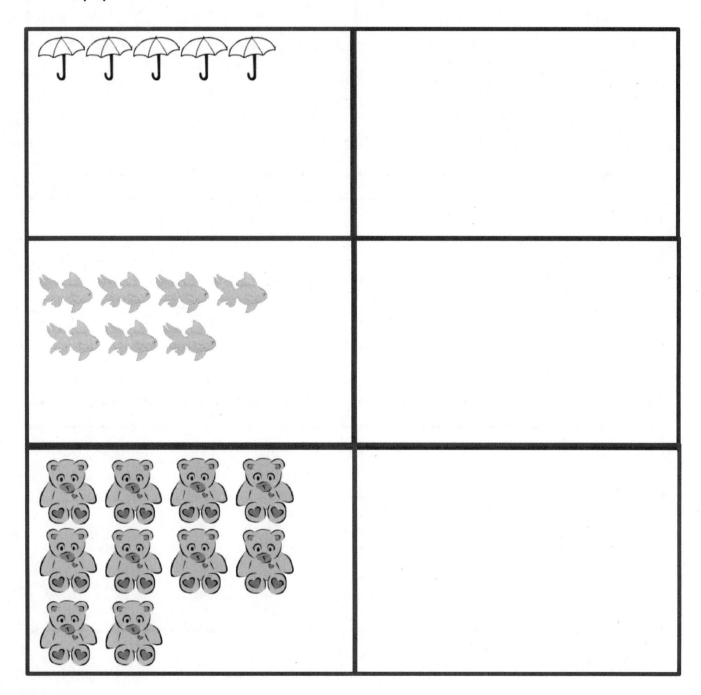

Lesson 22:   Identify and create a set that has the same number of objects.

109

©2015 Great Minds. eureka-math.org
GK-M3-SE-B2-1.3.1-01.2016

Draw a set of objects in the first box. Switch papers with a partner.
Have your partner draw the same number of objects in the next box.

EUREKA
MATH™

Name _____   Date _____

Count the birds. In the next box, draw the same number of nests as birds.

Count the houses. In the next box, draw the same number of trees as houses.

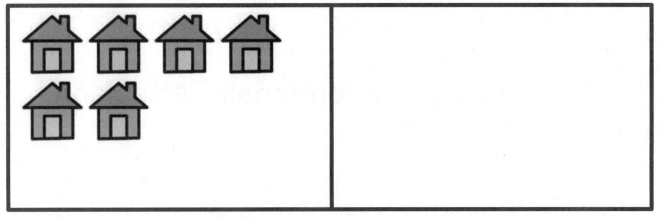

Count the monkeys. In the next box, draw the same number of bananas as monkeys.

On the back of your paper, draw some pencils. Then, draw a crayon for each pencil.

This page intentionally left blank

Name _____   Date _____

How many cats? [ ]

Draw a ball for every cat and 1 more ball.

How many balls? [ ]

How many elephants? [ ]

Draw a peanut for every elephant and 1 more peanut.

How many peanuts? [ ]

This page intentionally left blank

Name _____   Date _____

Count the set of objects, and write how many in the box.

Draw a set of circles that has 1 less, and write how many in the box. As you work, use your math words *less than*.

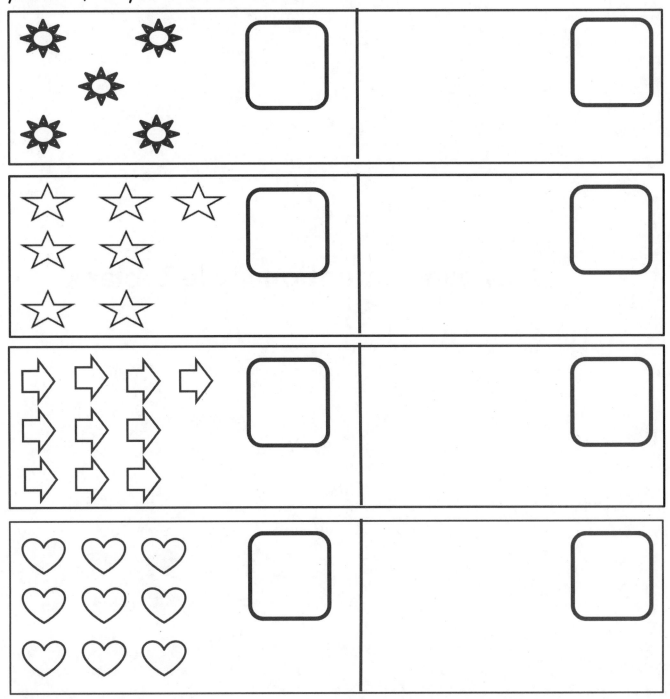

Lesson 24:   Reason to identify and make a set that has 1 less.

119

©2015 Great Minds. eureka-math.org
GK-M3-SE-B2-1.3.1-01.2016

This page intentionally left blank

Name _____   Date _____

Count the objects in each line. Write how many in the box. Then, fill in the blanks below. Use the words *more than* to compare the numbers.

_____ is more than _____.

_____ is more than _____.

_____ is more than _____.

Lesson 25:   Match and count to compare a number of objects. State which quantity is more.

121

©2015 Great Minds. eureka-math.org
GK-M3-SE-B2-1.3.1-01.2016

Roll a die, and draw a set of objects to match the number rolled. Write the number in the box. Roll the die again, and do the same in the next box. Use the words *more than* to compare the numbers.

_____ is more than _____.

_____ is more than _____.

_____ is more than _____.

    Lesson 25:   Match and count to compare a number of objects. State which quantity is more.

Name _____   Date _____

Count the objects in each line. Write how many in the box. Then, fill in the blanks below.

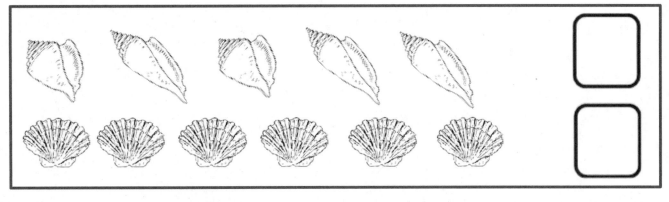

_____ is more than _____.

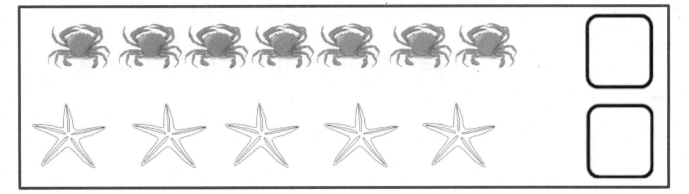

_____ is more than _____.

_____ is more than _____.

Lesson 25:   Match and count to compare a number of objects. State which quantity is more.

123

This page intentionally left blank

Name _____   Date _____

Count the objects in each line. _____ the blanks below. Say your w___aw a train with ____ewer cubes.

	Draw a tower with more cubes.
___ is less than ___.	___ is more than ___.

_____ is less than

_____ is less th___

___nother train with fewer cubes.

_____.   _____ is less than _____.

_____ is less t___

This page intentionally left blank

Name _____   Date _____

Draw a tower with more cubes.

_____ is more than _____.

Draw a tower with fewer cubes.

_____ is more than _____.

_____ is less than _____.

Draw a train with more cubes.

_____is more than _____.

_____is less than _____.

On the back, draw a tower. Draw another tower that has more cubes.

_____ is more than _____.   _____ is less than _____.

Lesson 27:   Strategize to compare two sets.

131

This page intentionally left blank

Name _____   Date _____

Visualize the number in Set A and Set B. Write the number in the sentences.

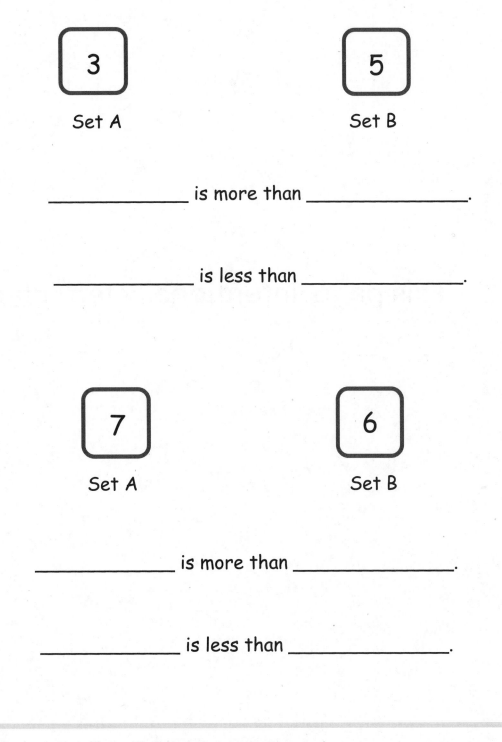

3

Set A

5

Set B

_____ is more than _____.

_____ is less than _____.

7

Set A

6

Set B

_____ is more than _____.

_____ is less than _____.

EUREKA
MATH™

Lesson 28:   Visualize quantities to compare two numerals.

133

©2015 Great Minds. eureka-math.org
GK-M3-SE-B2-1.3.1-01.2016

This page intentionally left blank

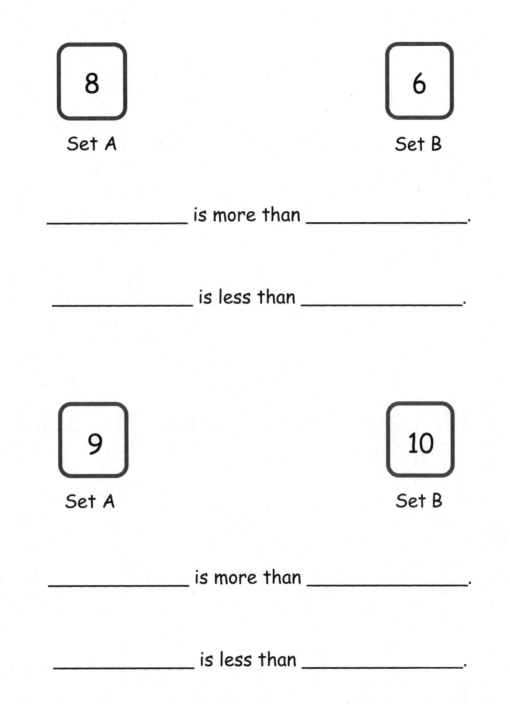

8

Set A

6

Set B

_____ is more than _____.

_____ is less than _____.

9

Set A

10

Set B

_____ is more than _____.

_____ is less than _____.

Roll a die twice, and write both numbers on the back. Circle the number that is more than the other.

This page intentionally left blank

Name _____   Date _____

Visualize the number in Set A and Set B. Write the number in the sentences.

7		4
Set A		Set B

_____ is more than _____.

_____ is less than _____.

9		10
Set A		Set B

_____ is more than _____.

_____ is less than _____.

This page intentionally left blank

8

Set A

6

Set B

_____ is more than _____.

_____ is less than _____.

4

Set A

5

Set B

_____ is more than _____.

_____ is less than _____.

Ask a family member to give you 2 numbers. Write the numbers on the back, and circle the number that is more than the other.

This page intentionally left blank

Name _____   Date _____

Draw a line from each container to the word that describes the amount of liquid the container is holding.

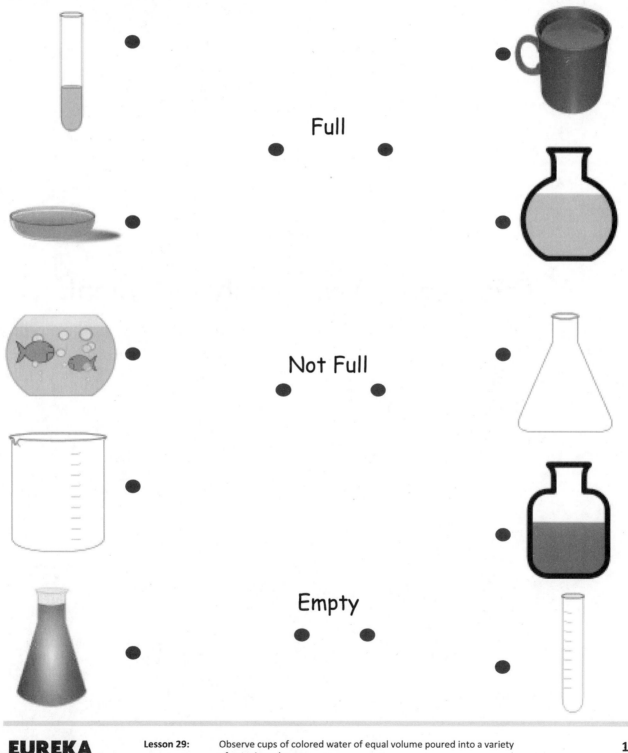

Full

Not Full

Empty

EUREKA MATH™

Lesson 29:   Observe cups of colored water of equal volume poured into a variety of container shapes.

©2015 Great Minds. eureka-math.org
GK-M3-SE-B2-1.3.1-01.2016

141

This page intentionally left blank

Name _____   Date _____

My Capacity Museum!

my capacity museum recording sheet

Lesson 29:   Observe cups of colored water of equal volume poured into a variety of container shapes.

143

©2015 Great Minds. eureka-math.org
GK-M3-SE-B2-1.3.1-01.2016

This page intentionally left blank

Dear Parents:

In class, we used balls of clay that weigh the same on the balance scale to make different sculptures. We saw that the same balls of clay can take various forms without changing the weight. The balls weighed the same on the balance scale, as did the sculptures.

Today's homework is a review of fluency work.

Name _____   Date _____

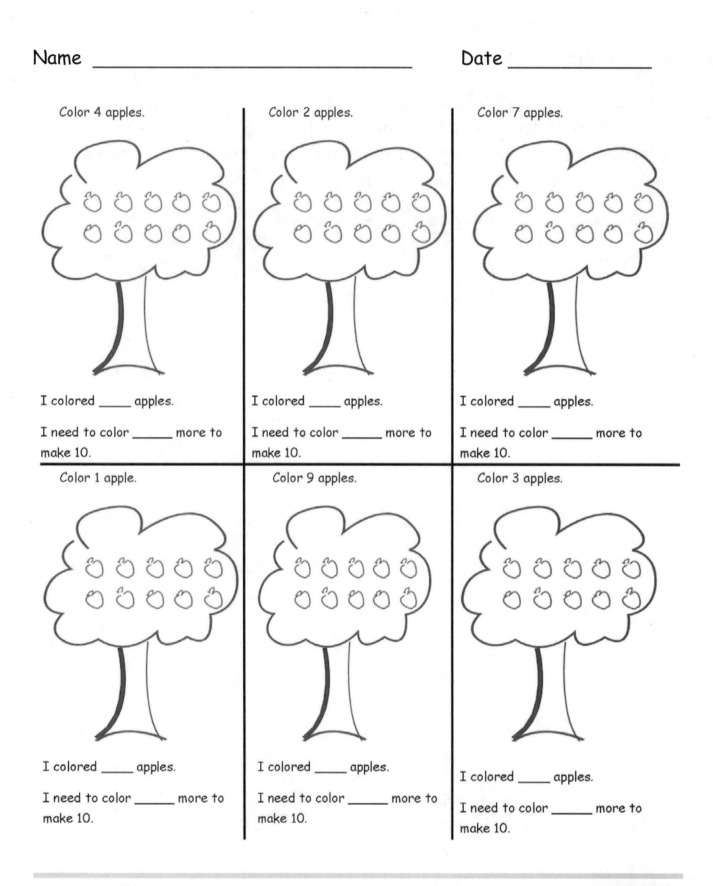

Color 4 apples.

I colored _____ apples.

I need to color _____ more to make 10.

Color 2 apples.

I colored _____ apples.

I need to color _____ more to make 10.

Color 7 apples.

I colored _____ apples.

I need to color _____ more to make 10.

Color 1 apple.

I colored _____ apples.

I need to color _____ more to make 10.

Color 9 apples.

I colored _____ apples.

I need to color _____ more to make 10.

Color 3 apples.

I colored _____ apples.

I need to color _____ more to make 10.

©2015 Great Minds. eureka-math.org
GK-M3-SE-B2-1.3.1-01.2016

Name _____   Date _____

Clay Shapes

clay shapes recording sheet

This page intentionally left blank

Name _____ Date _____

Listen to the directions, and draw the imaginary animal inside the box.

Draw a rectangle body as long as a 5-stick.
Draw 4 rectangle legs each as long as your thumb.
Draw a circle for a head as wide as your pinky.
Draw a line for a tail shorter than your pencil.
Draw in eyes, a nose, and a mouth.

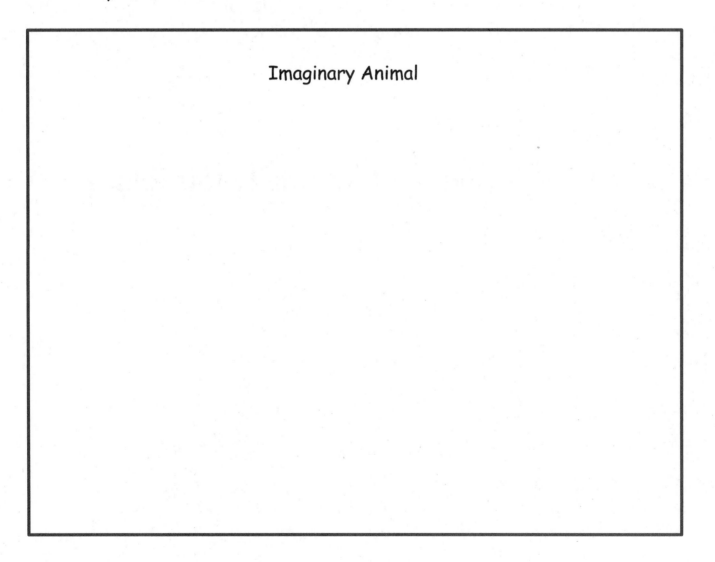

Imaginary Animal

Lesson 31: Use benchmarks to create and compare rectangles of different lengths to make a city.

149

©2015 Great Minds. eureka-math.org
GK-M3-SE-B2-1.3.1-01.2016

This page intentionally left blank

Name _____   Date _____

Read the following directions to your child to make a castle:

- Draw a rectangle as long as a spoon.
- Draw another rectangle on each side of the rectangle you just made.
- Draw a triangle on top of each rectangle to make towers shorter than your hand.
- Draw 1 rectangle flag pole as long as your pointer finger.
- Draw 1 square flag as long as your pinky.
- Draw a door as long as your thumb.
- Draw 2 hexagon windows each as long as a fingernail.
- Draw a prince or princess in your castle.

Castle

Lesson 31:   Use benchmarks to create and compare rectangles of different lengths
            to make a city.

151

©2015 Great Minds. eureka-math.org
GK-M3-SE-B2-1.3.1-01.2016

This page intentionally left blank

Name _____   Date _____

The homework is a review of fluency skills from Module 3.

Circle a group of dots. Then, fill in the blanks to make a number sentence.

__2__ and __4__ is __6__

_____ and _____ is _____.

_____ and _____ is _____.

_____ and _____ is _____.

_____ and _____ is _____.

On the back, make your own 6-dot cards. Circle some dots, and then say, "_____ and _____ is _____."

EUREKA MATH™

Lesson 32:   Culminating task—describe measurable attributes of single objects.

153

This page intentionally left blank

Name _____   Date _____

comparing attributes recording sheet

This page intentionally left blank